What Is It?

By Leslie Kimmelman

CELEBRATION PRESS

Pearson Learning Group

rough

smooth

soft

What do objects feel like?

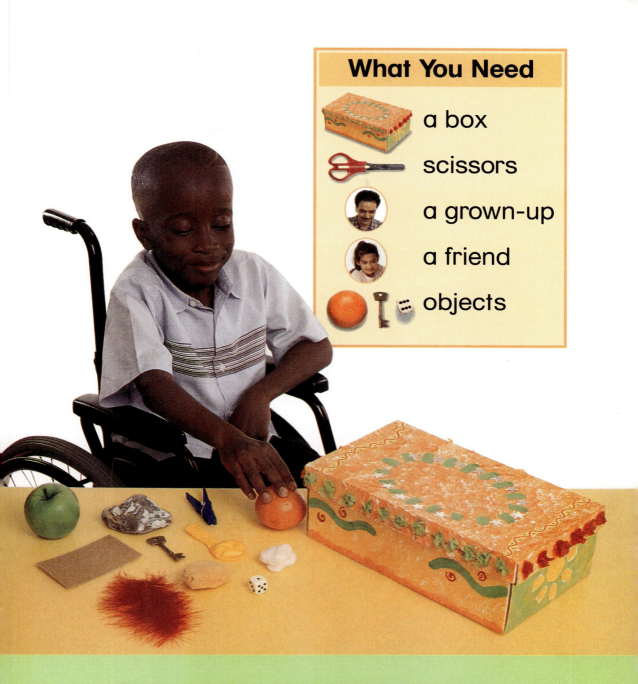

What You Need

a box

scissors

a grown-up

a friend

objects

Play a game to find out.

What to Do

1 Cut a hole in the box.

2 Put one object in the box.

5

3 Feel the object.
How does it feel?

4 Tell what the object is.

5 Play again with another object.